FND AND ME

By Mark Grimsley

Published by Mary River Press Services in Australia

Copyright © 2022

Find me @facebook.com/PTHRMM

@markgrimsleyphotos.com.au/

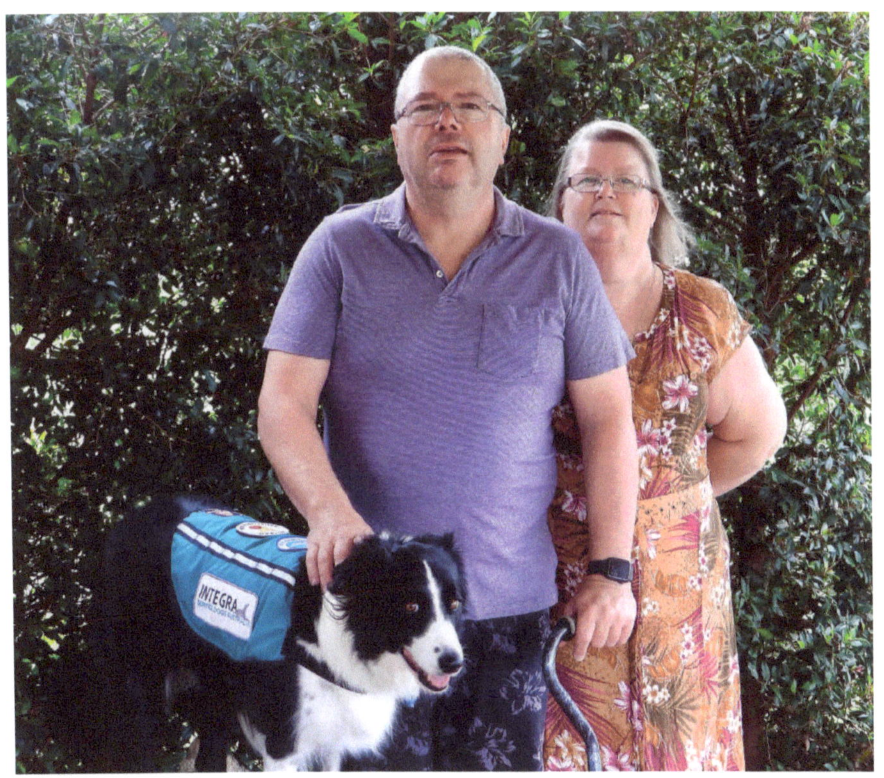

My name is Mark Grimsley, and I live on the Fraser Coast. I developed Post Traumatic Stress Disorder and Functional Neurological Disorder (FND) following an incident during deployment on active service as an aircraft life support fitter for the air force. I felt trapped by my disability until my daughter gave me a camera for Christmas, and I taught myself photography. This experience helped me reshape my mind and find some independence and connection with the community again. I now have an extensive collection of local bird and wildlife images. I wish to share my experience with my disorder and educate others about FND and service dogs. My wife, Kelly, and I formed a group called *FND group meet-up* that meets at the Hervey Bay RSL.

Did you know the second most common reason for a referral to a neurologist is FND, the first being migraines?

Functional Neurological Disorder (FND) is a medical disorder that disrupts the nervous system's function and how the brain and body send and receive signals.

Leg weakness, paralysis, seizures, walking difficulty, spasms, twitching, sensory abnormalities, and cognitive problems are some of the neurological symptoms associated with FND.

The core wiring of the nervous system is intact in FND, but there is an issue with how it functions and how the brain fails to send and receive messages accurately.

FND is a condition at the intersection of neurology and psychiatry, and the public and the medical profession rarely understand it.

I often fall, shake, experience seizures, and feel confused. My service dog helps get me through each day. When I fall, my border collie lies beside me and helps push me up when I am ready by burrowing her nose under my armpit.

My dog intuitively knows when I am about to shake and become disoriented, so she takes me to sit somewhere quiet.

I rely on my dog to keep me safe and want people to understand when working dogs are out in public, they have a job to do and must not be distracted.

TABLE OF CONTENTS

THE HORSES

I saw the horses on a day out on my trike. I was coming back along the beach, and the horses appeared from nowhere. They were on the beach at first and then saw me and went for a decent run out on the flats - a gallop - doing their own thing. They stopped twice to check my dog and me out. My dog, in the trailer, was as curious as I about the horses. The horses were looking back. It was an example of you never knowing what you will discover there! This image was one of my early images that drew feedback from people. They said they loved the image and the thought of the horses running wild on the beach. The feedback was amazing! People said they told their friends about my horse images, and when I bumped into their friends, they wanted to see the images. This built my self-esteem and confidence. It also allowed me to interact with people in a way I could not do previously. It was all because people were showing an interest in my images.

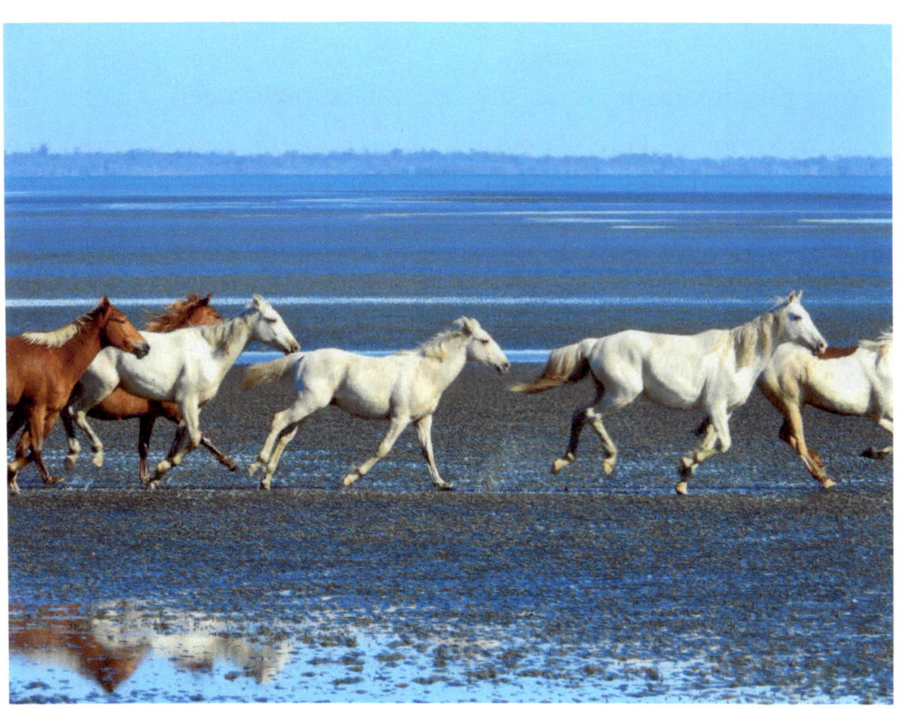

RAINBOW BEE-EATERS

I love this image because the bee-eaters have been a favourite since we moved here. They are always at the beach, and I see them most days. They are interesting because they nest into the ground for about a metre. On this day, I watched about half a dozen bee-eaters fly around and one perched on the log. As I was preparing to take the picture, another perched beside that one. I sat in my wheelchair, feeling very lucky to be photobombed, and it lifted my spirit.

I looked at the image later and noticed the log they were sitting on just seemed to pop right out and make the picture a little more special. Kelly and I loved the image so much that we printed one out on a glass image and gave it to a friend who was moving away. It looked terrific exhibited like that, and it is still one of my favourite images.

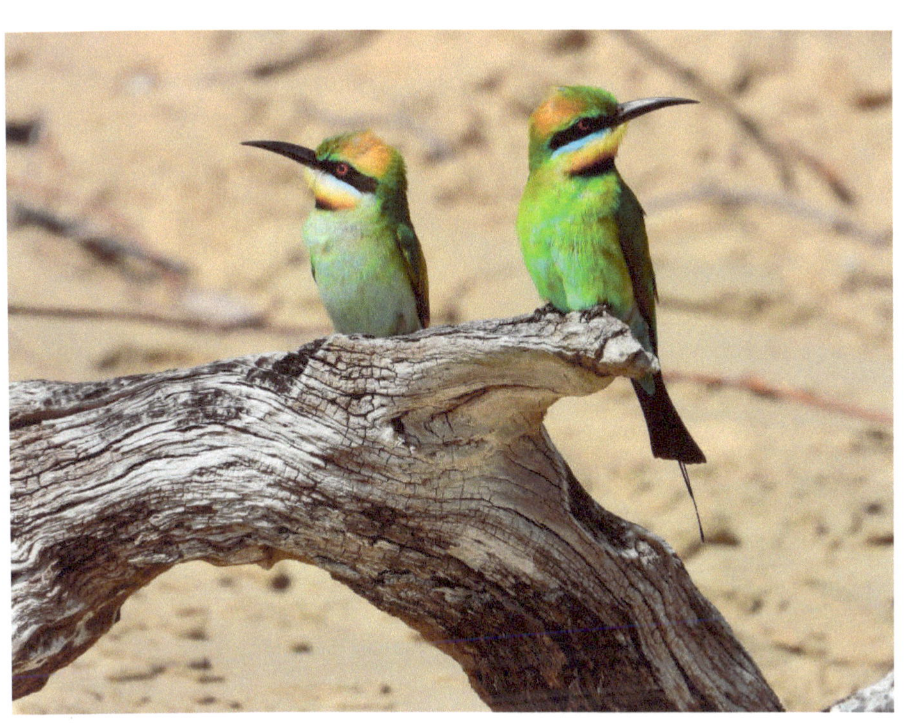

TORTOISE WITH A DRAGONFLY

It was around nine in the morning, and I was travelling around the lagoons. I do the lagoon circuit when the tide is too high for me to get on the beach with my wheelchair. A high tide restricts my access. This image was taken around the lagoons. I was trying to get a shot of the turtle's head reflected in the water, and as I pushed the shutter down, the dragonfly just appeared. It was a nice accidental shot. After I released the shutter, the dragonfly was gone. I have taken a photo of a tortoise with a dragonfly a couple of times since, but they were planned shots. The joy of this image was that it was a surprise.

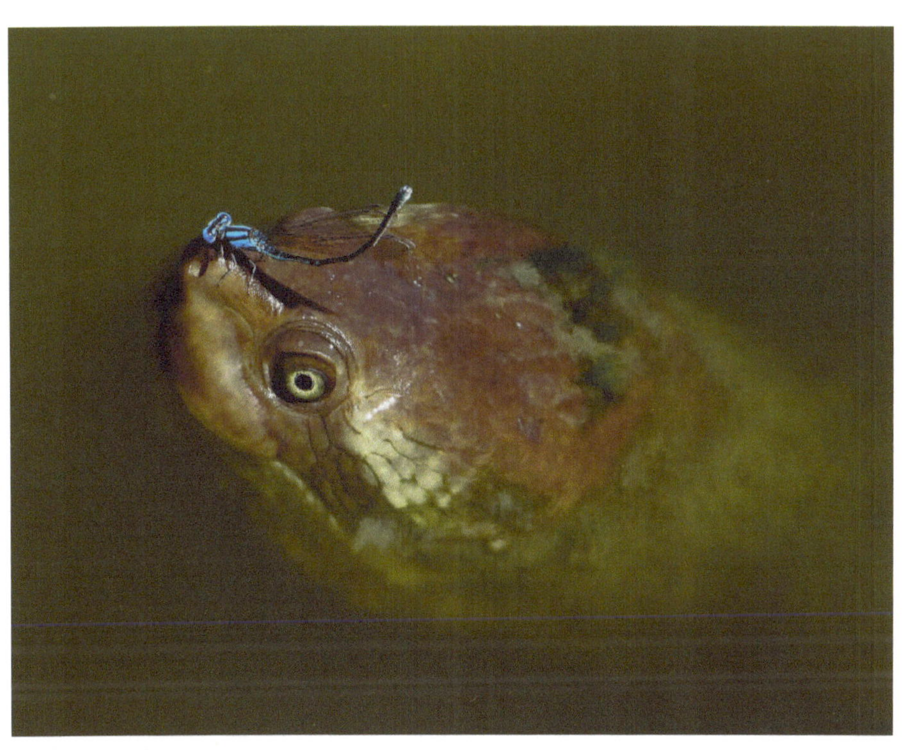

JABIRU

As my wife, Kelly, and I travelled in the car to the physiotherapist, I saw this Jabiru near a water hole. I was excited because it was the first time I had seen one. We went to the physiotherapist, and about forty minutes later, on the way home, it was still there! We raced home and got the camera. He was polite enough to be there when we returned about ten minutes later. I observed his beautiful display and assumed he was looking for food. He was stretching and walking around on the water's edge. Now, every time we go past that water hole, I look. We have seen him twice since then. I was probably there for about 25 minutes getting shots of him. These shots were taken from far away, over a road, through a barb wire fence, and in the fields. The camera my daughter gave me has such a good reach! I was stunned that day; I had never seen a Jabiru before. We go past that area most days, and because of this experience, I look there every time I go past. We see brolgas there as well at various times.

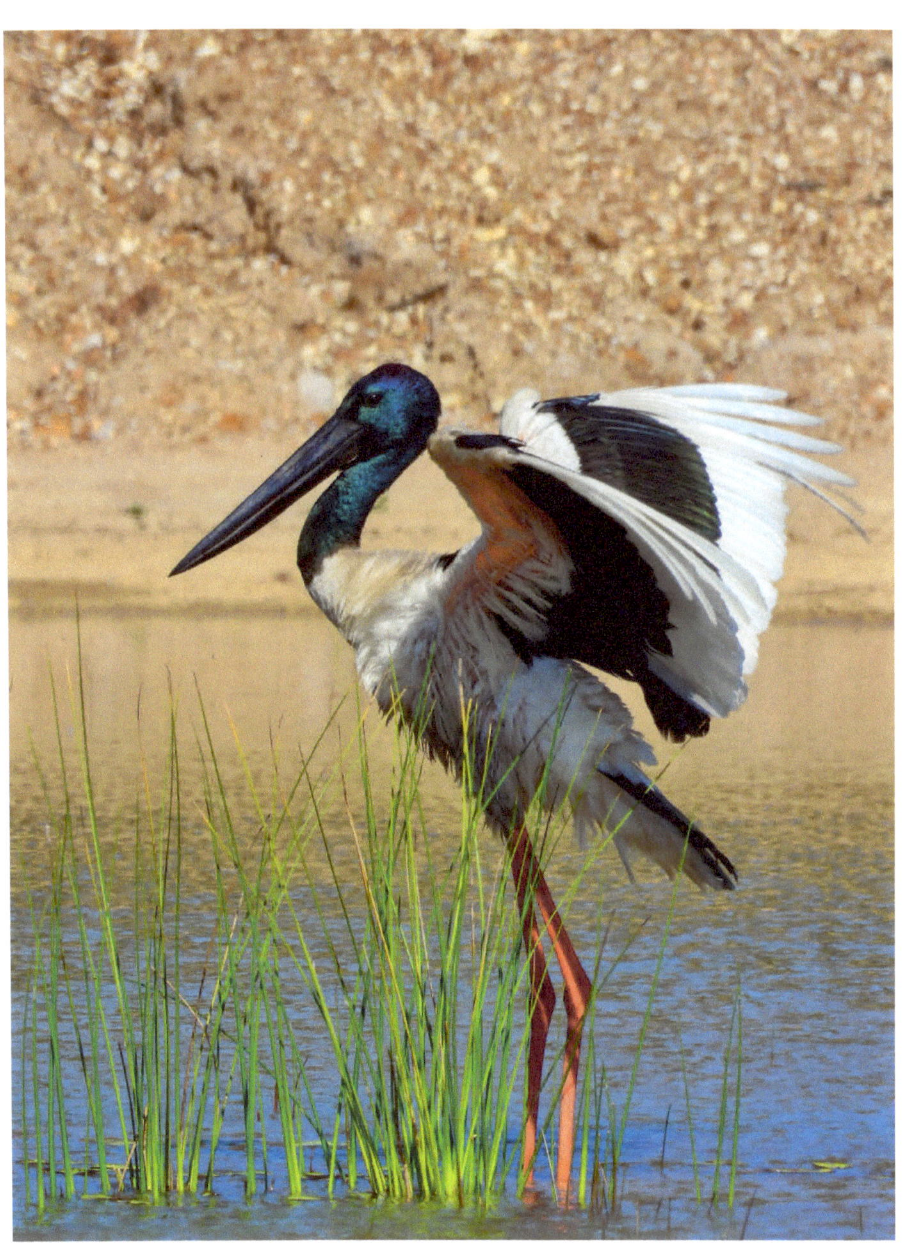

PHEASANT COUCAL

This image is what I call a WOW shot. This pheasant coucal was hanging around a big grevillea tree in my backyard and jumped down to one of the bird baths. After a while, he flew down into the back yard wandered around for about ten minutes. That was another wonderful experience because it was the first time I had seen one of those birds. I took the pictures while standing on the back patio. I was so pleased because it is why we put bird baths in - to attract different bird life.

When birds come, it is peaceful and calming, and I feel like they trust me enough to come into the backyard. Of course, we do not feed them anything, but we did plant 200 native trees to attract them, which has been a joyful project. The trees include bottle brushes, grevillea, hedges, bamboo, mulberry, blueberry and mango trees.

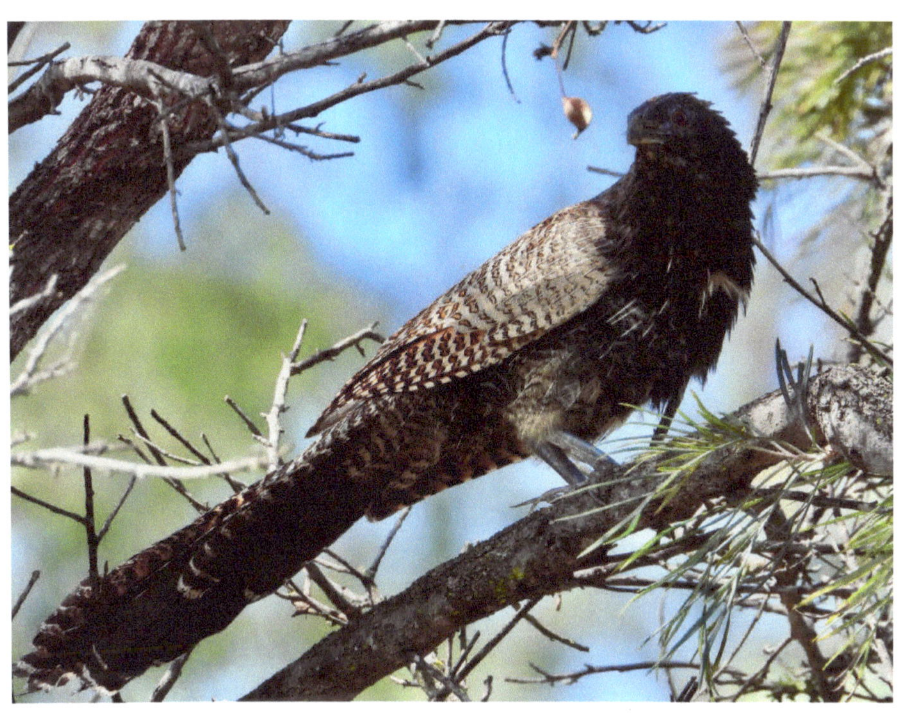

EDITED FLOWER

During the recent floods, we had over 133 millimetres on a Friday. We were lucky enough not to experience flooding. However, I cannot take the wheelchair out in the rain because the electrics will get wet – it will stop and will not go again. I also cannot take the trike because I wear the camera on my chest, and the camera will get wet. So, I was stuck at home and was going stir crazy. I felt I needed to get out and do something with the camera. When I don't, it messes with my thinking. Photography gives me a purpose and when I can't do it, I struggle because it feels like I have done nothing for the day.

I had already been cooped up for several days with the rain and couldn't get out on the wheelchair or the trike. I had to do something, so I grabbed the umbrella, the camera and a torch and walked around the garden to figure out what images I could take in the wet. By shining the torch on a part of it and including the raindrops, I created a shot and was happy with the result. Eight and a half thousand people saw the images! Even though taking photos of flowers was outside my comfort zone, and I only did it because of the weather conditions, these photos were the best I've ever taken. Combined with Apple photo editing software, I created stunning images. My joy was reinforced by lots of lovely feedback on Facebook about them being beautiful and creative. The whole experience was a challenge and different from wildlife images because I had to piece them together.

After taking the images, I was happy with the results and was inspired to look for more ideas. It kept me occupied, and I went out a couple of times that day to try again with mushrooms and others that were not quite as successful. I found that by being creative with my photography, I could challenge myself, and it kept my thinking straight.

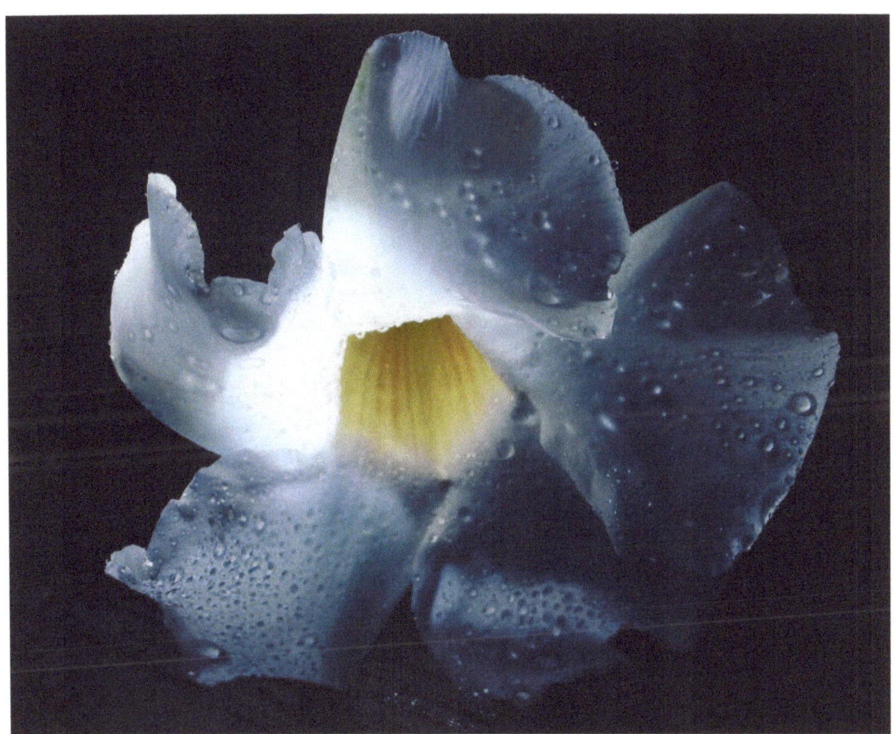

PALE-HEADED ROSELLA

This image was luck; I took the picture after being out for a couple of hours with limited success in getting shots. There was nothing super out there that day, so I returned to my street. A rosella was flittering around the bush. I saw movement and the camera helped me get nice and close. I quickly tried to grab a couple of shots. The rosella didn't want to hold still and kept moving. I didn't think I got anything significant. Out of what I took, I got one image I liked. I was happy to get one, at least.

As I usually do, I posted it on my Facebook page, and I was surprised other people liked that shot and delighted that someone purchased that single image.

I have tried to get rosellas before and can't get clear images. So I was glad to get this one.

It was a wonderful experience because I was switched off; I thought the photography was finished, and I was heading home. The beauty of photography is that I don't overthink it, so many of my shots are spontaneous.

I was pleasantly surprised and excited that the shot was so popular. It was an inspiration to keep going. Even when you are about to put the camera away, there might still be an opportunity to take a photo.

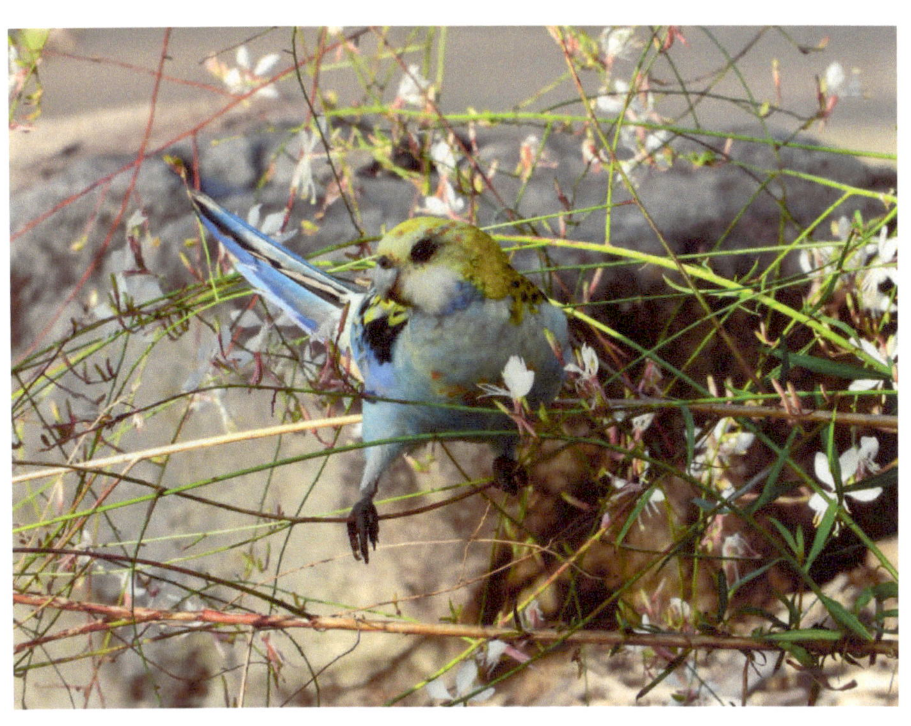

THE TREE

I've been taking photos of the same tree for about three years, and I love getting different shots of it. One morning I got a shot of the sun coming through a hole in the middle of the tree. There are a lot of dead trees along the beach, but this one is just hanging on and hanging on. I will be a little sad when the ocean takes this one away. There's been a couple that haven't lasted as long. I posted one of these shots on Facebook, and a lady commented that she loved the image, which was obviously photoshopped. I was surprised because it wasn't. I think she assumed that because the tree stood there and looked out of place. At the same time, it didn't look out of place.

I love riding up to it and taking shots. I will miss it when it goes because it is like a marker on the beach and a place to go to and part of my routine. Routine is essential to keep me on track. The military made me a routine person. My routine has helped me with my disorder. Riding up to the tree is an example of my routine, and I usually stop there for five or ten minutes and look around. Birds of prey are sometimes up at that end of the beach, and few people are there. I like to ride on my trike and see my tree.

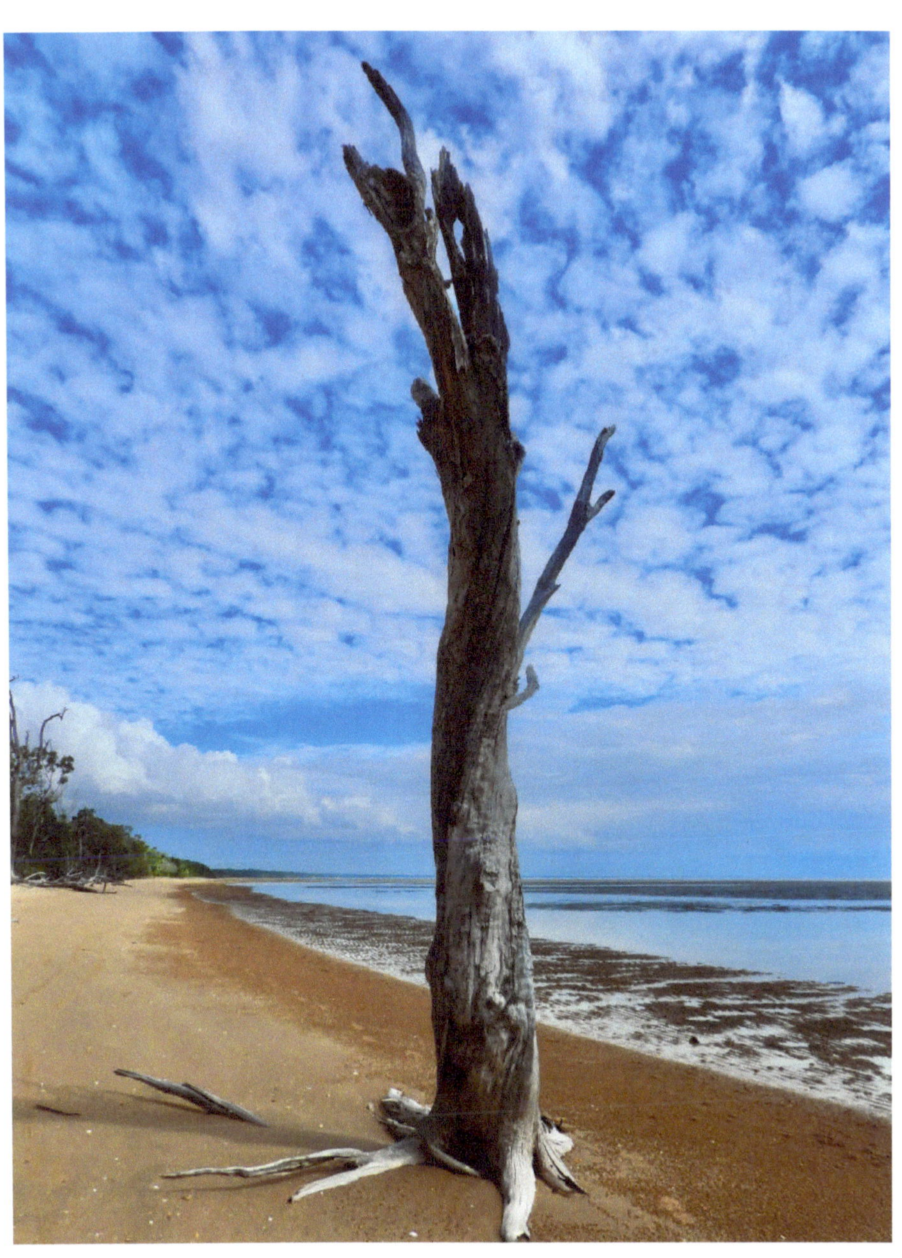

OSPREY

Before I had FND and PTSD, I was very athletic. Now, living with my disability, I feel constrained and can only walk around my house. To stop me from feeling stir crazy, I go to the beach on a purpose-built trike or motorised wheelchair. I had covered ten kilometres in my wheelchair on this day when I saw the osprey.

My favourite is the shot of an osprey high on a tree with a fish. A crow was bothering it, and the image I shot shows the osprey looking back, squawking at a crow with its wings spread out as it is swooping the osprey. I assume it wanted to take its food away. On several occasions, I have got shots of ospreys. On this day, several birds were hanging around, and I was trying to get an image of the dead wood. I think it adds something else to it. The whole image demonstrated how wonderful nature is, and even when you think the osprey has its food and that's what you will get a shot of, something else comes along and completely changes what you are looking at for that day.

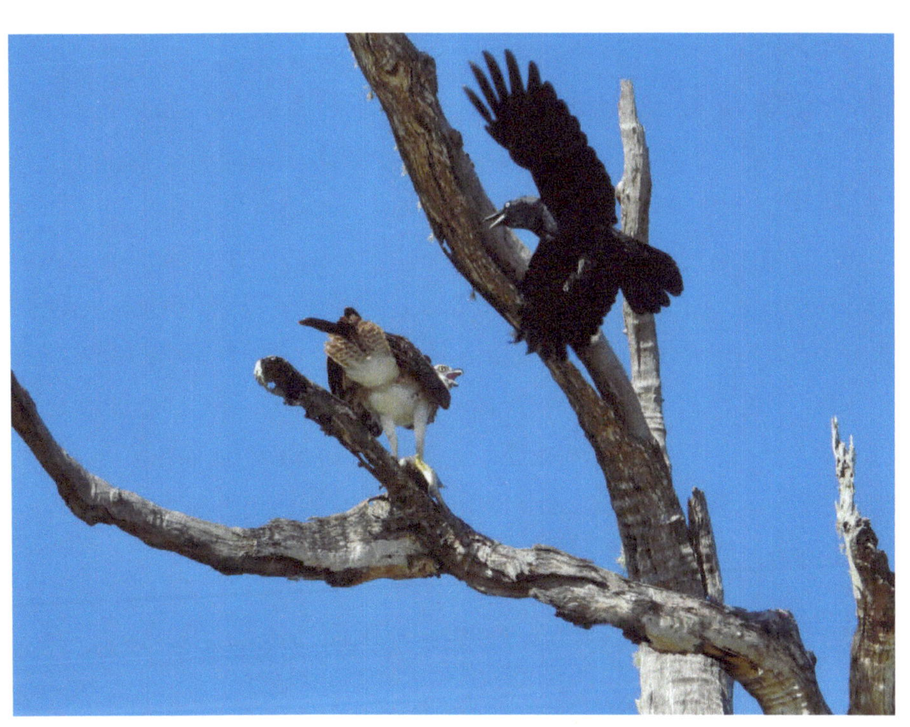

SEA EAGLE

I like the sea eagle because people have said they are hard to get close to because they are shy. I've ridden up the beach and found the sea eagle in a tree a few times. The eagle is not worried about me being there on the trike as it looks over the ocean for its next meal. Unfortunately, the trike can only access that far up the beach as it is too far for the electric wheelchair. I will be stuck on the beach if the wheelchair runs out of battery. The sea eagle is unique, and it is quite calming to see such a big bird sitting there, not bothered by you. He looks majestic yet fearsome, putting me on a high when I see him.

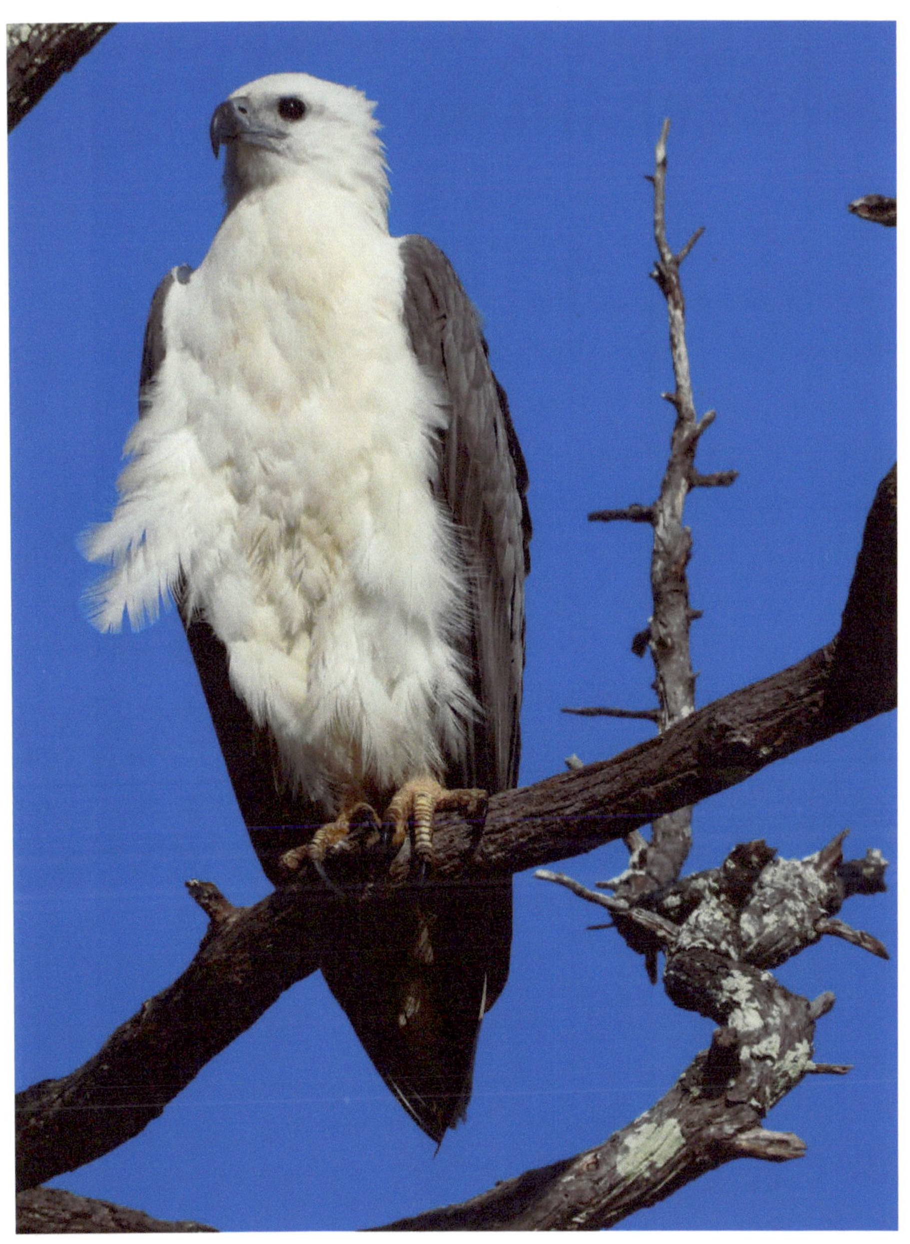

BRAHMINY KITE

The brahminy kite is why I wanted to take on photography in the first place. I was down on the beach four and a half years ago, and a couple of them were sitting on the sand. I had no idea what they were. I tried to take a picture with my phone, which was a terrible photo. I had no way of identifying what the birds were from the image. It sparked my initial interest in learning more about the local birds. And it started my interest in photography because I wanted better images of what was out there. Still, I get excited when I see the brahminy kites. A couple of times, they have shown curiosity and flown around my trike. Most of the time, they are quite happy to sit in their tree and let me get dozens of shots of them. They look so relaxed sitting up there. They make you excited.

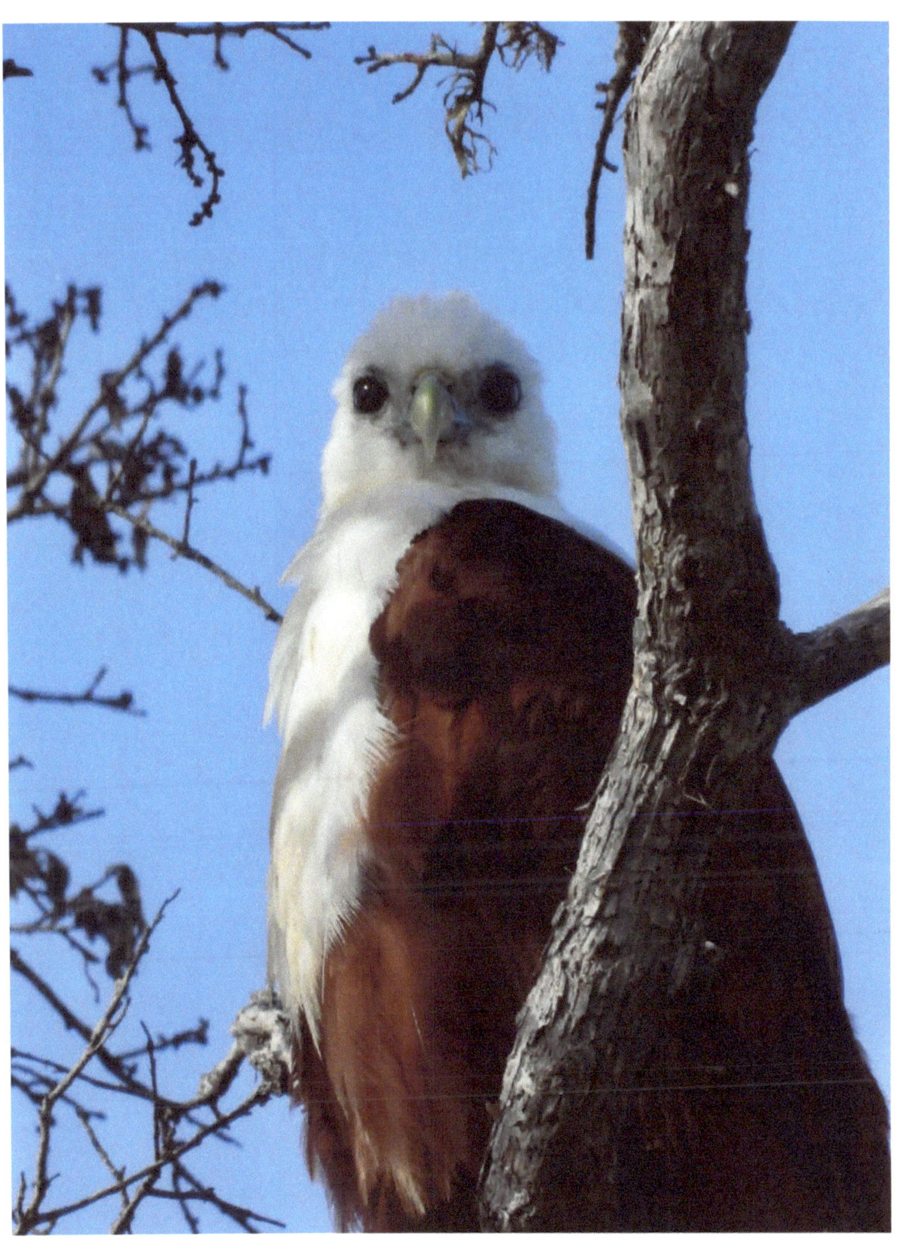

WHISTLING KITE

I've often tried to get shots of birds of prey in flight, and it isn't easy because you have to focus and get the movement quickly.

Down the beach, if they fly into a headwind, they will hang for a little while, making it easier for you to get them.

On this day, three or four kites flew around trying to chase a sea eagle. One got bored, flew a couple of hundred metres toward me, and circled me a few times. He was coming toward me, and they seem like they own the skies when they are out there. I pick up on the movement, which is usually not on their own. Most of the time, there is a group of them. They also go up to the isolated end of the beach where the sea eagle is. They seem a little timider than the other birds of prey down here.

I knew nothing about birds. I had only seen a bird of prey at a display at Taronga Zoo. And before picking up a camera, I had no idea how many birds of prey species we had on the beach here. Unfortunately, I still do not have a photo of everyone local in this area, so I look forward to finding more.

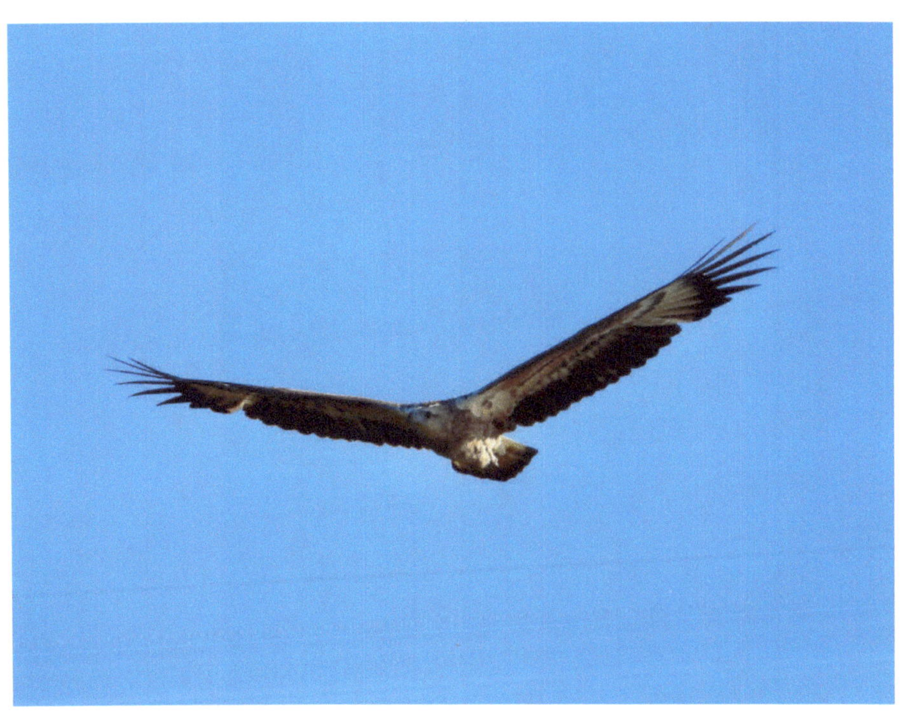

9 780645 112351

FND
AND ME

After developing Post Traumatic Stress Disorder (PTSD) and Functional Neurological Disorder (FND) following an incident during deployment on active service as an aircraft life support fitter for the air force, Mark felt trapped by his disability. One Christmas, his daughter gave him a camera, and he taught himself photography. This experience has given him hope, independence, and connection with the community. Mark shares his journey with images from his collection to help others find a way out of despair.

ISBN 978-0-6451123-5-1

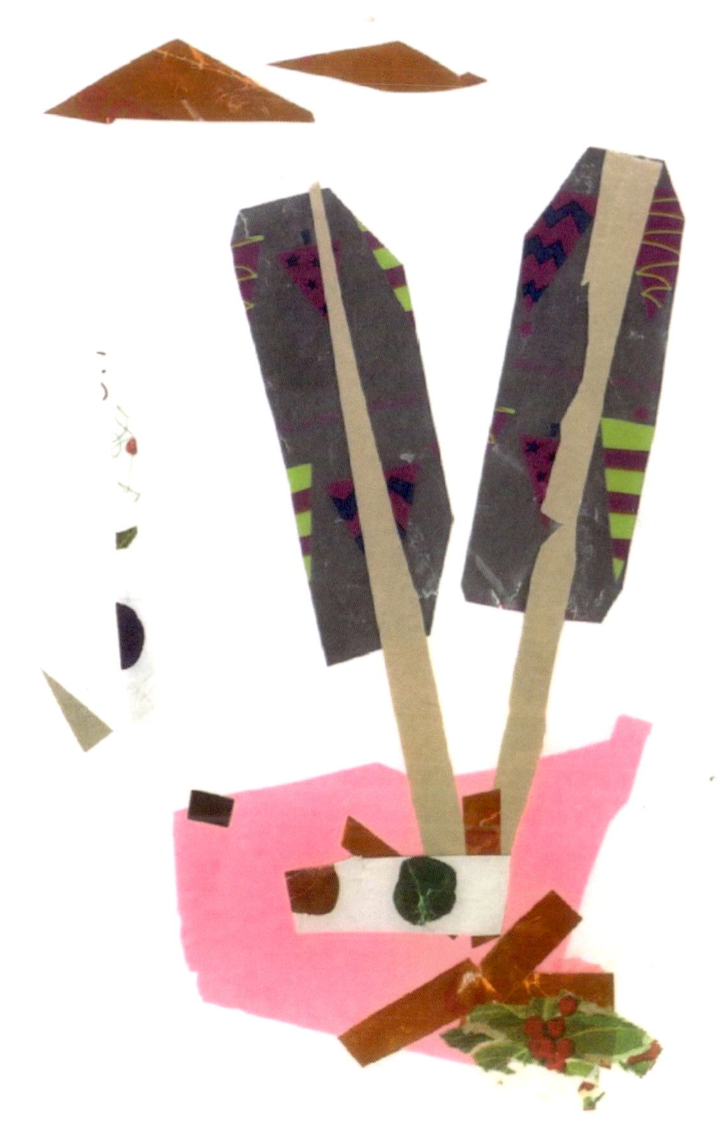

WRAPPED IN LOVE

Joe Pascoe